BODY TALK
DIGESTION
THE DIGESTIVE SYSTEM

JENNY BRYAN

Wayland

BODY TALK

BREATHING

DIGESTION

MIND AND MATTER

MOVEMENT

SMELL, TASTE AND TOUCH

SOUND AND VISION

THE PULSE OF LIFE

REPRODUCTION

Editor: Catherine Baxter
Series Design: Loraine Hayes
Consultant: Dr Tony Smith – Associate Editor of the *British Medical Journal*
Cover and title page: Silhouette of girl eating

First published in 1992 by Wayland (Publishers) Ltd.
© Copyright 1992 Wayland (Publishers) Ltd.

British Library Cataloguing in Publication Data
Bryan, Jenny
Digestion.– (Body Talk Series)
I. Title II. Series
612.3

ISBN 0 7502 0416 8

Typeset by Key Origination, 1 Commercial Road, Eastbourne
Printed in Italy by G. Canale & C.S.p.A., Turin

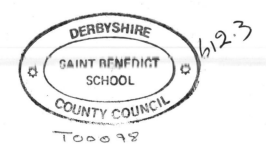

CONTENTS

Introduction	4
Where your dinner went	6
The unsung heroes	8
A balanced diet	10
A healthy weight	12
Dieting sensibly	14
Appetite	16
Malnutrition	18
Anorexia and bulimia	20
Digestive problems	22
Food allergies and intolerance	24
Diabetes	26
Upset stomachs	28
An irritable bowel	30
The great beef scare	32
Cancer	34
The dangers of alcohol	36
Poisoning	38
Sweet or savoury?	40
Tooth decay	42
Future food	44
Glossary	46
Books to read	47
Index	48

INTRODUCTION

Cornflakes, chocolate, crisps, cheese, curry, cola, coffee – these are just some of the things that the digestive system may have to put up with in a normal working day. It's a long day too. Hours after we have gone to bed our intestines are still hard at work.

The digestive system was designed to cope with raw food long before humans discovered how to make fire and cook. It could get nutrients from raw meat, leaves and fruit. But, today, we expect it to digest rich, spicy, processed foods without a murmur of complaint.

We expect too much. Our digestive system has found out how to rebel. Diarrhoea, vomiting, constipation, indigestion, ulcers and, if we ignore all these warnings, cancer, are some of the ways in which our guts fight back.

But we don't have to spend a lifetime doing battle with our digestive systems. If we treat them carefully and eat the right kinds of food, there will be no complaints. That doesn't mean we have to give up all our favourite foods and eat only what is good for us. But it does mean we should stop eating so much 'junk food'.

RIGHT The amount of fresh vegetables we eat hasn't changed much in the last thirty years, despite all the publicity about how good they are for us. But we do eat more fresh fruit and we drink nearly twice as much fresh fruit juice as in the early 1960s. Far more people are now vegetarian (they do not eat meat).

LEFT If you put these foods in front of someone who lived in the late 1800s they wouldn't even know what they were! Our diet has changed dramatically during the twentieth century. Many of us could not even imagine what life was like before 'fast foods' took over.

WHERE YOUR DINNER WENT

What did you have for dinner yesterday? Let's assume it was an average sort of meal, say, meat, Brussels sprouts, carrots and potatoes, followed by apple pie and cream. Perhaps you had a fizzy drink to wash it down.

Where did it go? First your teeth broke up each mouthful of food into amounts small enough to go down your throat. From there it went down the tube which leads from your mouth to your stomach. This is called the oesophagus.

In the stomach, your dinner churned round and round. Hydrochloric acid produced by cells in the wall of the stomach started to turn your food into a semi-liquid mush. Enzymes started to break down the proteins in your meal, mainly found in the meat.

After an hour or so, your dinner was ready to go down into the small intestine – the top part is called the duodenum. There, different enzymes started to break down the carbohydrate in your meal, mainly in the potato and the pastry of the apple pie. Enzymes also started digesting the fat, mainly in the meat and the cream.

All the carbohydrates, fats and proteins have to be broken down into simple molecules before they can be absorbed through the wall of the duodenum and, later, the wall of the ileum. The ileum is the second, much longer part of the small intestine. It is 5 m long but it takes up much less space than you'd think. That's because it is coiled up inside your belly.

The inner wall of the small intestine isn't smooth. It has thousands of tiny folds. This gives it a much larger surface area than if it were smooth. So more nutrients can be absorbed from the intestine into the bloodstream. It is here that vitamin A would have been absorbed from the carrots in your meal and vitamin C from the sprouts and the apple.

It takes several hours for food to get down the small intestine and pass into the large intestine (colon), through the caecum. The large intestine is the packaging service for waste products from your meal that aren't absorbed in the small intestine. The waste products are the fibre from the sprouts, carrots and apple and any dead cells mucus, and unused chemicals. Your small intestine needed the fibre to push the food through although it didn't absorb it.

Fluid is absorbed from the colon. So that's where the fizzy drink would have parted company with the remains of your meal.

The caecum and colon are about 1.5 m long and lead into the much shorter rectum which is only 13 cm long. In the rectum, waste in the form of faeces is stored ready to be excreted from the anus – the opening at the end of the intestines. This could be as much as forty-eight hours after you swallowed your first mouthful of dinner.

Which foods would you cut out if you were trying to improve her diet?

THE DIGESTIVE SYSTEM

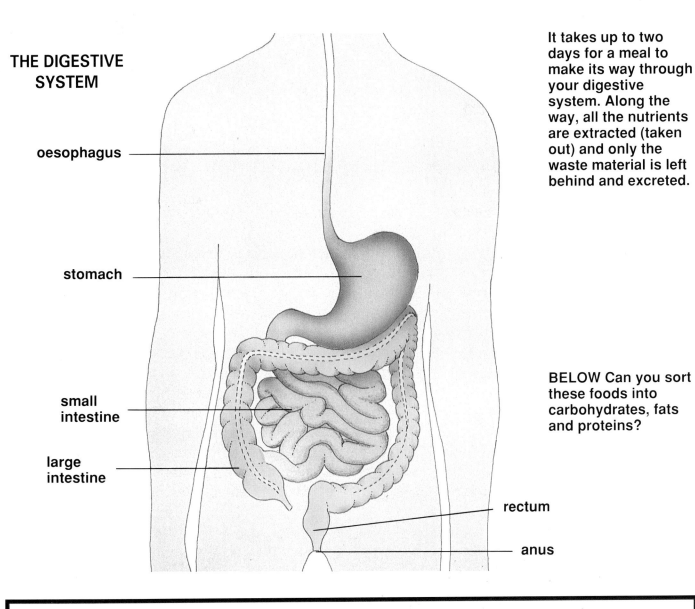

- oesophagus
- stomach
- small intestine
- large intestine
- rectum
- anus

It takes up to two days for a meal to make its way through your digestive system. Along the way, all the nutrients are extracted (taken out) and only the waste material is left behind and excreted.

BELOW Can you sort these foods into carbohydrates, fats and proteins?

CARBOHYDRATES, FATS AND PROTEINS

Nearly all our food is carbohydrate, fat or protein. Only vitamins and minerals are different. Carbohydrates are found in the starchy and sugary foods in our diet, such as bread, potatoes, pasta and sweet drinks. Fats are found mainly in dairy products, such as milk, cream and butter. Proteins are found in meat, fish and, in smaller amounts, vegetables and fruit.

After they have been digested and absorbed into the bloodstream, carbohydrates and fats are used for energy. Proteins are broken down into molecules called amino acids that are needed for growth and repair in the body.

THE UNSUNG HEROES

The enzymes that are needed to digest carbohydrates, fats and proteins in the intestines do not appear by magic. They are made or stored in the liver, the gall bladder and the pancreas.

The liver is the largest organ in the body and lies just below the diaphragm – the sheet of muscle that separates the chest from the abdomen. If you can feel your liver there's something wrong with it! A nice healthy liver is soft and slippery under your skin. It only becomes hard and shrivelled when it is diseased. That's when you can feel it.

The liver is probably the busiest organ in the body. It has several hundred different jobs. But two of them are especially important. Liver cells produce bile – a greenish-yellow fluid that contains salts needed to break down fats. The liver also removes toxins and poisons from digested food so they can be excreted.

Bile is released from the liver and stored in the gall bladder nearby. Inside the gall bladder the bile becomes stronger. During digestion, bile leaves the gall bladder and enters the small intestine through a small duct (tube).

The pancreas is a gland that lies alongside one of the loops of the small intestine. A number of enzymes needed for carbohydrate, fat and protein digestion are produced in the cells of the pancreas. They get into the duodenum through a special tube called the pancreatic duct.

The pancreas has another important role. A group of cells within the pancreas called the Islets of Langerhans produce a hormone called insulin. Insulin is needed to control the amount of glucose travelling around the body in the blood. People who don't produce enough insulin develop a disorder called diabetes. They are called diabetics.

Look closely and you can see where this diseased liver has become hard and shrivelled.

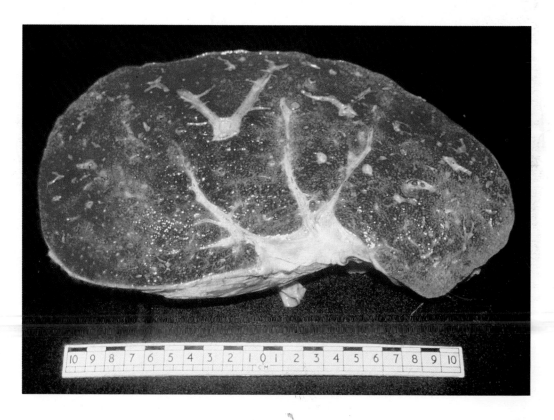

A single Islet of Langerhans from a human pancreas.

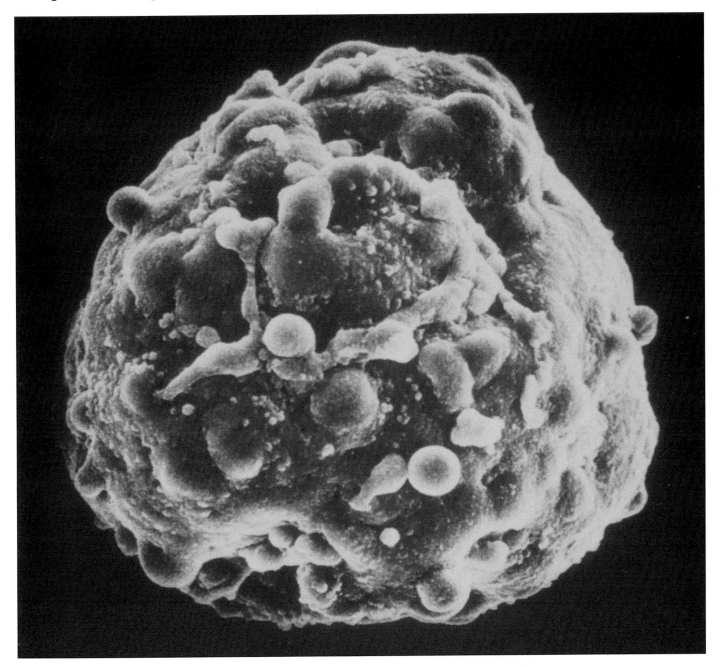

LIVER AND PANCREATIC TRANSPLANTS

These are less common than heart and kidney transplants. This is because liver transplants are very difficult to do and pancreatic transplants have never been very successful.

The liver has a very complicated blood supply which makes it hard to transplant. But at least you don't need to transplant a whole liver. Quite a small piece is enough to perform all the normal jobs of the liver.

Similarly, surgeons rarely transplant the whole pancreas. They have been experimenting with transplanting just some cells from the Islets of Langerhans for people with diabetes. But the results have been poor.

A BALANCED DIET

Eating a balanced diet isn't difficult. You just need the right mix of starchy and fatty foods, proteins, vitamins and minerals. If you eat a balanced diet you are more likely to stay healthy than someone who doesn't.

You need food to give you energy. A teenage girl should eat foods that will give her about 2,300 calories of energy per day. A teenage boy should get around 2,800 calories per day. Starchy foods such as bread, potatoes, rice and pasta provide a lot of energy. You can also get energy from sweet things such as cakes and biscuits. But sweet foods are bad for your teeth and often contain a lot of fat.

Fat is also a source of energy because it is high in calories. But your body cannot handle large amounts of fat. If you eat a lot of fatty foods you will put on weight and fat will probably start to build up in the walls of your arteries. This may eventually lead to a heart attack.

It's much healthier to get most of your calories from foods that do not contain a lot of fat or sugar.

No one expects you to go without fat altogether. But you can limit the amount by drinking skimmed and semi-skimmed milk, adding yoghurt instead of cream to desserts and using polyunsaturated margarine instead of butter. You can even get 'low fat' margarines and cheeses.

In a balanced diet, you also need protein. Most people get this from meat and fish. White meat, such as chicken, contains less fat than red meat, such as beef or lamb. So try to eat more chicken than red meat.

Vegetarians don't eat meat. But they can get the protein they need from vegetables. These contain less protein than meat so it is important for a vegetarian to eat other foods that contain protein, such as soya beans, cereals, nuts, and fruit if they are to stay healthy.

We should all eat plenty of vegetables and fruit for a balanced diet. These contain fibre to help push food through the intestines. They also contain vitamins and minerals.

DO WE NEED VITAMIN AND MINERAL SUPPLEMENTS?

Not if we eat a balanced diet containing at least one vegetable and one piece of fruit every day. Vitamin and mineral supplements should never be used instead of eating vegetables and fruit. They are less effective because the body cannot handle them as well as 'real' food.

However, elderly people and people with long-term illnesses may benefit from supplements. This is because they may not be able to eat enough of the right sort of food to get the necessary nutrients to keep their bodies working normally.

Some people believe that extra vitamins and minerals can protect them against a wide range of illnesses from colds to cancer. But there is very little scientific evidence to support this.

How many of these grains and pulses have you eaten? They include lentils (orange), Aduki beans (green), red kidney beans, black-eyed beans and chickpeas.

A HEALTHY WEIGHT

When they are in their early teens most girls are taller and weigh more than boys. This is because girls reach puberty one to two years before boys. An average eleven-year-old girl weighs between 27 and 45 kilograms and an eleven-year-old boy about the same. At fourteen, the same girl will weigh between 45 and 65 kilograms and a fourteen-year-old boy between 40 and 60 kilograms.

People who are very overweight are damaging their health. They are more likely to suffer from heart disease, breathing problems, diabetes, arthritis and some forms of cancer. They are also likely to get very depressed and miserable about their weight. Losing weight is much harder than putting it on! So you should try very hard to stay the right weight for your height.

Don't blame your parents if you are fat. Children whose parents are very tall or short do tend to take after them. But just because your parents are a bit plump, it doesn't mean that you have to be overweight too. It does mean that you may have to be more careful about what you eat. That means being more strict about the amount of biscuits and sweets that you eat and choosing fresh fruit juices instead of sugary fizzy drinks.

If you get into the habit of healthy eating when you are young, you will save yourself years of miserable dieting when you are older.

There are no hard and fast rules as to how much we should weigh. It changes according to our height, general build, our sex, and whether or not we have reached puberty.

THE REAL HEAVYWEIGHTS!

Most sumo wrestlers die young because they are so overweight.

An average sumo wrestler from Japan weighs about 130 kilograms. He consumes 7,000 calories a day and lives until he is about forty-five. During his professional life he uses his sheer bulk to push his opponents out of the sumo wrestling ring. But carrying all that extra weight does untold damage to his heart and other organs. So he dies at least twenty-five years before the average man.

DIETING SENSIBLY

There is only one way to lose weight. That is to break down your excess fat and use it to give you the energy you need.

In theory, there are two ways of doing this. You can eat less so your cells cannot get enough energy from the calories in your food. In response your body will start to raid your fat stores. Or, you can increase your body's need for energy by increasing the amount of exercise you take. Since you won't be able to get any more energy from your food, your body will need to attack your fat.

In practice, you need to take a great deal of exercise to lose weight – more than most of us can fit into a normal day. So, to lose weight, you will have to reduce your energy (calorie) intake and that means going on a diet. (Taking more exercise as well as dieting will help to speed up the rate at which you lose weight.)

A sensible way to lose weight is to cut down the amount of fat you eat and to exercise. You need to be very active to use up large amounts of calories. If you can find an activity you enjoy you are more likely to stick at it for longer and really feel the benefits.

If you go on a diet it's not enough just to eat less. A small bar of chocolate takes up less space on a plate than a large helping of salad. But it has a lot more calories! It's just as important to eat a balanced diet when you are trying to lose weight as when you are eating normally. However, you should aim to cut out fatty or sweet foods because these contain the most calories.

Many people can lose weight when they go on a diet. But they find it very hard to stay slim when they come off their diet. This is because they are very strict with themselves for a few weeks or months. Then they start eating the foods again that made them fat in the first place.

Before you go on a diet look at what you eat now and decide what has made you put on weight. Do you eat a lot of crisps, sweets, biscuits, cakes or fried food, or do you snack between meals? You'll have to cut some of these out even when you have finished your diet.

The chocolate is more enticing, but it's much more fattening than the lettuce.

HOW TO LOSE WEIGHT

Decide how much weight you want to lose and how long it should take. Be realistic! You should not expect to lose more than one kilogram a week. Most diets cut your energy intake to about 1,000 calories per day. If you go below that you'll probably find it too hard to stick to the diet.

Agree a series of meals for a week with your parents which will enable you to keep to 1,000 calories a day. Don't skip meals but be very strict with yourself about eating between meals. A few simple rules are: eat fish and chicken rather than red meat; eat more salads, vegetables and fruit; don't eat sweets or biscuits; eat grilled or steamed food, not fried; swap from dairy products to low calorie milk and spreads. Check the number of calories on the packaging of the food you eat. You'll be surprised at some of the figures!

Crisps may taste really nice, but they contain a lot of calories and additives.

APPETITE

How often do you eat because you are really hungry? Be honest! Don't you sometimes eat because you like something rather than because you are hungry?

Humans weren't always like that. All animals have evolved with systems to control their eating habits. These controls are in the brain. There are two centres – a hunger centre which tells you when to eat because you need energy, and a satiety centre which tells you when to stop eating because you have had enough. But humans – more than any other species – have learned to override the messages from these centres. So they can eat for pleasure as well as to satisfy the body's needs. This is why you see more overweight people than overweight cats or dogs!

If we listened only to the messages from our hunger and satiety centres, we would eat only when the levels of glucose in our blood were low. Glucose is the most important substance which our cells need to create energy. Fat and carbohydrates from food are broken down to form glucose.

Soon after a meal the amount of glucose in the blood is high. This stimulates the satiety centre to make you feel full. But when the body has used up most of the glucose in the blood the hunger centre tells us that it is time to eat again.

It isn't just greed that can make us override the centres in the brain. If you feel sick or ill, you may not feel like eating even if the glucose level in your blood is quite low and your brain is sending out messages that it is time to eat.

Human beings have learnt to override the messages from their satiety centres. This means that we can go on eating long after we have had enough. Here, comedian John Cleese offers the fattest man in the world (played by Terry Jones) a last mouthful after he has eaten a light snack of everything on the menu in 'Monty Python's Meaning of Life'.

Here, a Muslim family break their fast with dates and water during Ramadan. During Ramadan, Muslims do not eat or drink anything for a month between sunrise and sunset.

FASTING

Every year, during Ramadan, practising Muslims fast from sunrise to sunset. This is ordered by the *Qur'an*, the Muslim holy book. For a month, no food can pass their lips during the day. No one pretends it is easy to ignore the feelings of hunger during that time. But fasting for a day – or even a few days – is unlikely to do any harm to someone who is in good health to begin with.

The first eight hours of a fast are probably the worst. As you become more hungry, you find yourself continually thinking about food. After a few days you feel less hungry, but you may find it increasingly hard to concentrate.

You can go without food for longer than you can go without water. After only a few days without water you would become extremely dehydrated and risk serious organ damage. So you should not fast and go without water too.

MALNUTRITION

LEFT The stick-like arms and legs, and the dull and hopeless eyes, tell an all-too-familiar story of hunger and malnutrition for this child from Guatemala.

RIGHT This child has rickets. She didn't get enough vitamin D. As a result her bones softened and her wrist became misshapen.

Malnutrition means that, over a long period of time, someone is not getting the sort of food that he or she needs to stay healthy. It is generally used to describe what people in developing countries are suffering from when they cannot get enough food. You do also occasionally hear of cases of malnutrition in Europe.

Each day, millions of people in Africa, Asia and South America do not get enough to eat. Some are painfully thin. They have used up all the stores of fat in their bodies and they are starting to break down muscle. If they have a big stomach, it is not because it is full of food, but because it is swollen by disease.

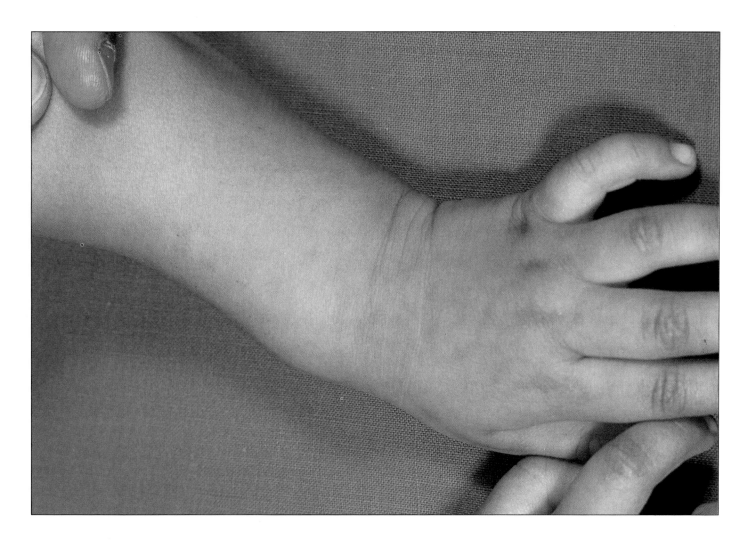

THE EFFECTS OF FAMINE

Malnutrition can cause permanent damage. Even if a starving child is given more food and starts to put on weight, it may suffer from permanent brain and other organ damage. People with malnutrition are also more likely to get infections and they are less able to fight them off. Even coughs and colds, mumps and measles can be fatal.

If you don't get enough food, you don't just miss out on the fats and carbohydrates you need for energy, and protein for growth. You don't get enough vitamins.

Vitamins are essential for good health. Deficiencies can lead to many different diseases. Serious vitamin A deficiency can lead to blindness. Vitamin B1 deficiency can cause beriberi – nerve and muscle damage. Vitamin B12 deficiency leads to anaemia – a blood disorder. Vitamin C deficiency causes scurvy – a painful skin condition – and vitamin D deficiency results in rickets.

Worldwide, we produce enough food to feed everyone. But 950 million people in the world are undernourished and at least 400 million get less than 80 per cent of their needs. Meanwhile, people in rich countries eat on average 30-40 per cent more calories than they need.

Children always come off worst. Right now, four out of ten children under five in developing countries are suffering from malnutrition. Each year over fourteen million children worldwide die before they reach their fifth birthday. Malnutrition plays a part in a third of these deaths.

ANOREXIA AND BULIMIA

Some people go on strict diets even when they are not overweight. They eat so little that they become dangerously thin. Without help they can die. They have anorexia nervosa – or slimmer's disease. Anorexia is most common among girls in their mid teens. But teenage boys can get it, as can children as young as eight or nine.

People with anorexia have nothing wrong with the appetite centre in their brains. They deliberately suppress their desire to eat. Most have emotional problems that they have been unable to sort out. Refusing to eat is their way of reacting to what is upsetting them. Obviously, it is a very bad way.

Some anorexics are very unhappy with themselves. They feel they are worthless as people. Others do not get on with their families or they cannot cope with growing up and becoming more independent. Few anorexics think they are thin. They look at themselves in the mirror and still see a fat person. Their idea of fat and thin is totally distorted. When they are told they must put on weight they are horrified.

Some anorexics are so thin by the time they see a doctor that, unless they are forced to eat, they will die. They have to go into hospital and medical staff are very strict with them about eating. It isn't very nice. But it is necessary to prevent anorexics from starving themselves to death.

When they are out of danger, doctors try to find out the emotional problems that have led to the anorexia. This usually means talking to the whole family and probably helping to change the way they all feel about each other. Fortunately, most people with anorexia do get better. They usually stay rather thin. But they stop trying to diet themselves to death.

This anorexic woman thinks that she is fat. But her thin arms and legs tell a different story.

RIGHT Someone with anorexia would never eat a slice of cake. Bulimics on the other hand would eat the lot and soon after make themselves sick.

BELOW Many anorexics and bulimics take large quantities of laxatives so that they don't digest their food properly.

Bulimia nervosa is also an eating disorder. But people with this disease are not usually as thin as those with anorexia. They may eat huge quantities of food in just a few minutes. This is called bingeing. But they then feel guilty and make themselves sick so they bring it all back up. This is clearly very bad for them. Again, there is usually an emotional cause for this behaviour and successful treatment depends on finding out what the problem is.

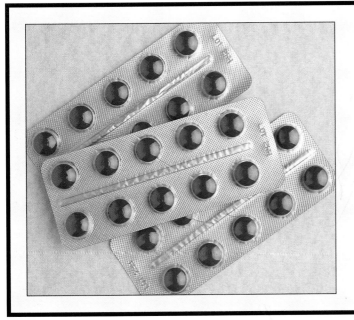

HELPING SOMEONE WITH ANOREXIA OR BULIMIA

If you think you know someone with anorexia or bulimia it is very important that you encourage them to get help. There are all sorts of signs. Anorexics and bulimics can be very cunning. They may seem to be eating properly, but they may hide their food and throw it away or rush to the toilet after a meal to make themselves sick. Many take a lot of laxatives so they do not digest their food properly and have continuous diarrhoea.

All of this is very dangerous. Try to convince them to seek help. Or, if they won't listen, talk to your parents or to a teacher about it. You won't be 'telling tales'. You may be saving their life.

DIGESTIVE PROBLEMS

Everyone gets indigestion sometimes. Indigestion is your stomach's way of complaining about the way you are mistreating it. You're most likely to get it if you eat too much or too quickly. If you 'wolf' down your lunch and rush straight out to play games you are asking for trouble!

You will probably burp and feel rather uncomfortable. You may get a burning feeling in your chest. People call this heartburn. If you ignore the symptoms and go on rushing about you may even be sick.

Some people get indigestion even when they eat sensibly. This is because they have something wrong with their stomachs. The cells in the walls of their stomachs make too much acid. This can damage the delicate lining of the stomach and cause an ulcer which may even bleed. The ulcer may be in the stomach itself or in the duodenum.

You probably know what it's like to have an ulcer in your mouth. It's very sore and tender. Imagine something similar in the stomach and remember that, each time you eat, there'll be more acid running over the ulcer.

Until the mid 1970s someone with an ulcer usually needed an operation. Surgeons would cut the nerves to the stomach that told the cells to make acid. This meant that no acid could be produced. The ulcer healed and the pain went away, but there was a risk of permanent problems with digestion.

Don't go swimming after a big meal. It may make you sick.

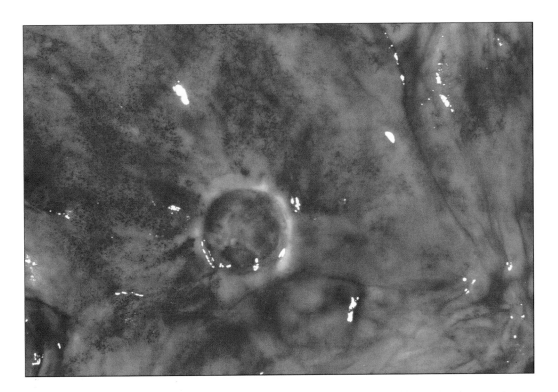

You can see why stomach ulcers are so painful. They are red and inflamed and they sometimes bleed.

In 1976 a drug became available that could reduce the amount of acid produced by the stomach but did not block it altogether. The patient simply took the drug every day for a few months. With less acid in the stomach the ulcer healed, the pain went away and the drug could then be stopped. It was so easy! If the ulcer came back, the patient could have another course of the drug.

Since the 1970s a number of similar drugs – all called histamine (H2) antagonists – have been made. They work by preventing histamine – a substance in the body – from triggering acid production in the stomach cells.

HP – THE MYSTERY BUG

In the last few years doctors have discovered that people who are prone to recurrent stomach or duodenal ulcers have a bacterium in their digestive system called *helicobacter pylori*. No one is quite sure how it gets there or why it gives people ulcers. But if people get rid of the bacterium with antibiotics, ulcers are less likely to come back.

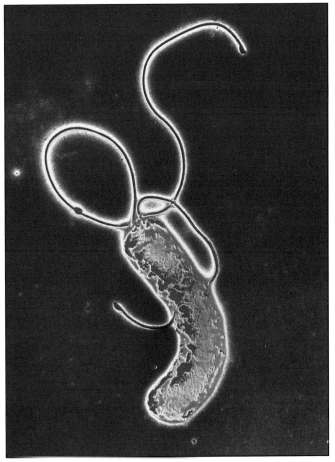

This is *helicobacter pylori* – the bacterium which has been linked to stomach ulcers.

FOOD ALLERGIES AND INTOLERANCE

One in five people think they are allergic to certain foods. They may be sick after they eat them or get a blotchy rash on their faces or necks. But strictly speaking, few have a true food allergy.

A food allergy – like allergies to dust, pollen or animal fur – must involve the immune system. When the sufferer swallows the offending food, special proteins, called antibodies, are produced in the blood. These antibodies call up large supplies of white blood cells and they all attack the food. In doing so they cause soreness, swelling and redness in the skin, and they may upset the stomach too.

Many people react badly to foods, but their immune system is not involved. They have a food intolerance rather than a food allergy. A food intolerance can occur because someone is missing an enzyme or some other chemical needed to break down a particular food. The symptoms can be just as unpleasant as an allergic reaction.

People are allergic to all sorts of things – not just food. Hay fever sufferers, who are allergic to pollen, would probably sneeze and itch if they were anywhere near this field.

You can tell the difference between food allergy and intolerance. If you are ill the very first time you have a particular type of food, it's more likely to be food intolerance. This is because your immune system simply cannot react quickly enough the first time you eat something you are allergic to. You only get an allergic reaction when you eat the food for a second time.

Dairy products, nuts and seafood are the most common causes of food allergies. The tiniest amounts can set some people off.

In contrast, someone with a food intolerance can often get away with eating a small amount of the offending food and only get a reaction when they eat a lot of it.

People who have other allergic illnesses, such as asthma or hay fever, are more prone to food allergies. But food allergies can go away. So, unless you had a really violent reaction to something, it's worth trying it again every few years. For example, many children are allergic to milk when they are small. But they often grow out of it.

Some doctors try to 'desensitize' people to the foods they are allergic to. By giving someone a small amount of the 'allergic' food every day, they try to make their immune system used to it so they stop over-reacting. But this is very difficult and rarely works. If you know that you have a food allergy it's best to avoid the food that causes it.

A lot of people are allergic to dairy products.

MIGRAINE
The chemicals in certain foods can trigger migraine headaches in people who are sensitive to them. Red wine, chocolate, and oranges are the main culprits. A migraine is a severe headache usually accompanied by disturbed vision and/or disturbed hearing, and sickness. As with food allergies, it is best to avoid the foods that cause migraine.

DIABETES

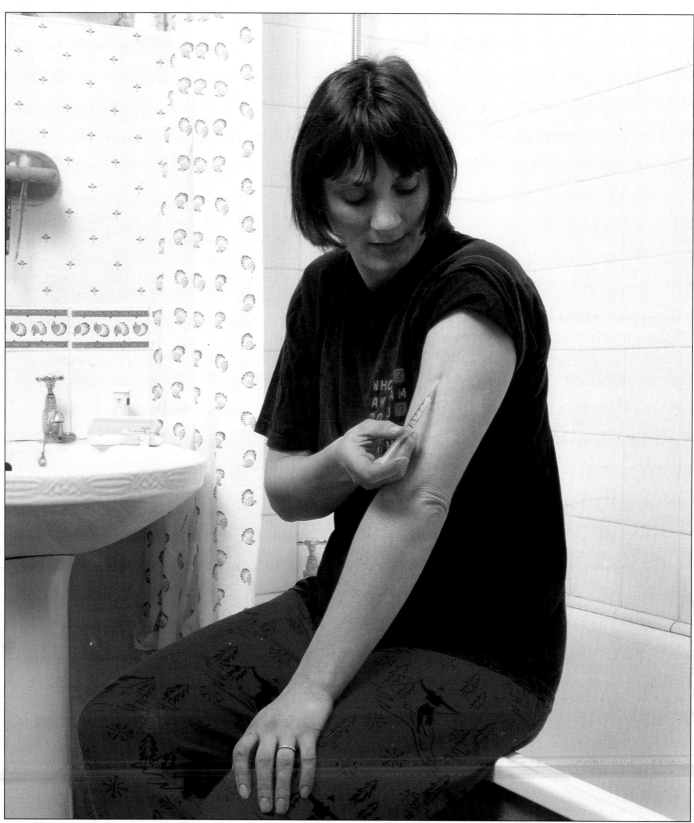

A diabetic woman injects herself in the arm with insulin.

About 2 per cent of people in industrialized countries have a disease called diabetes. It means they cannot control the amount of glucose in their blood because they do not make enough of a hormone called insulin. Some people are born with diabetes. The cells in the pancreas that normally produce insulin do not work. Other people get diabetes when they are older. They still produce some insulin, but not enough.

If diabetes is not treated, the amount of glucose in the blood goes up and up. This causes serious organ damage. People can lose consciousness and they may die.

Fortunately, diabetes can be treated very effectively. People who are born with the disease can give themselves daily injections of insulin to control their glucose levels. They can measure the amount of glucose in their urine or blood to make sure their treatment is working.

People who develop diabetes in later life may not need insulin injections. They may be able to control their glucose levels by cutting out sweet foods from their diet and by taking tablets that help their pancreas to make more insulin.

It isn't possible to prevent babies from being born with diabetes. But it is possible to reduce your risk of getting it when you are older. People who are overweight are more likely to get diabetes – another good reason to watch your weight.

It is very important for people with diabetes to take their illness seriously and make sure their glucose levels are kept under control. If they don't, they are more likely to get eye, kidney, nerve and foot problems when they are older.

HELPING SOMEONE WHO IS HAVING A 'HYPO'

Sometimes a diabetic's glucose level falls too low. This is called a 'hypo'. Usually they feel weak and faint. The quickest way of increasing the amount of sugar in their blood is to get them some sweets or a biscuit.

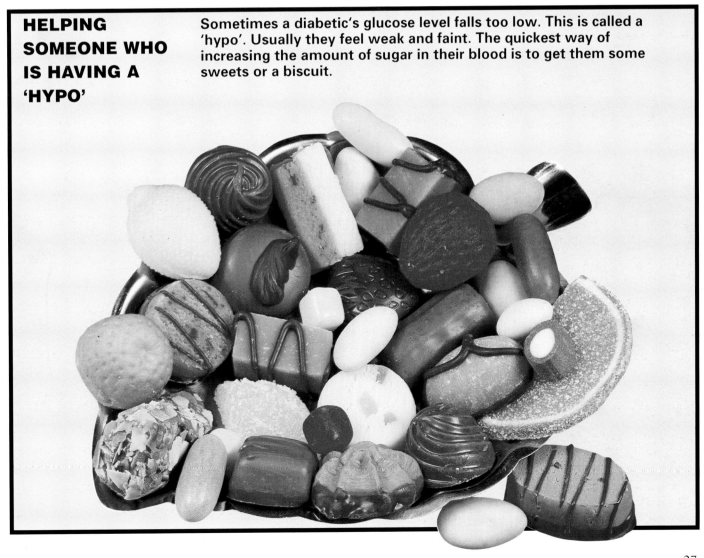

UPSET STOMACHS

Food poisoning is more common than ever before. Each year, thousands of people have to take time off work or school because they ate food that was contaminated by micro-organisms.

Bacteria are most often to blame. The best known is *salmonella* – a family of bacteria that is found in raw meat, eggs and unpasteurized milk. Another bacterium to hit the headlines in the last few years is *listeria* which is found in soft cheese and paté.

People who travel to developing countries, where standards of hygiene may be poor, risk other types of infection which are carried in food. These include typhoid, caused by a bacterium, and hepatitis A, which is caused by a virus.

Common symptoms of food poisoning include diarrhoea, vomiting and stomach pain. Sufferers may also have a high temperature. It may be difficult to work out exactly what has poisoned you as symptoms can occur anything from a few hours to several days after you were poisoned. Sometimes it is the bacteria themselves that cause the trouble. Or the bacteria may produce poisonous toxins that make you ill.

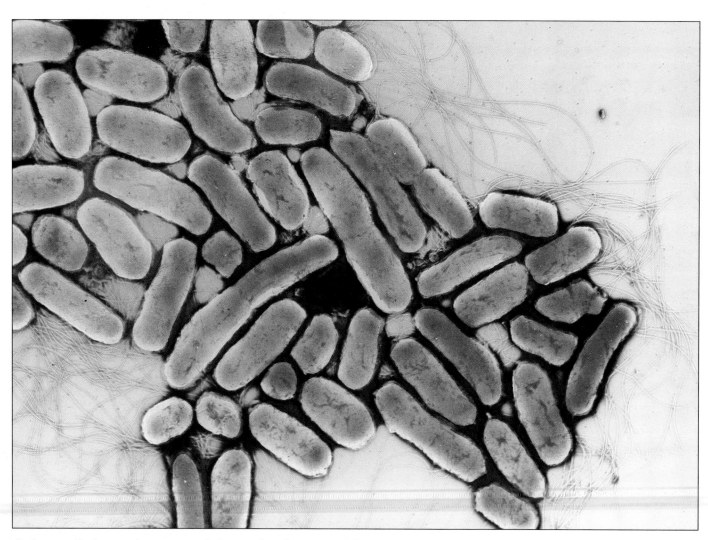

Salmonella bacteria causes thousands of cases of food poisoning each year. There was a particularly severe outbreak in Britain in 1988.

RIGHT Food poisoning frequently attacks in the middle of the night. As well as being sick you may also have diarrhoea. It's easy to become dehydrated so make sure that you have plenty to drink.

BELOW RIGHT Babies, young children, pregnant women, ill people and the elderly should not eat eggs when the yolk is still runny.

The best way to treat most cases of food poisoning is to let nature take its course. It's important to drink plenty of fluids to replace those lost by vomiting and diarrhoea as it's easy to become dehydrated. Drinks should have added sugar and salt to help restore the body's natural balance of chemicals.

SAFETY FIRST

We can protect ourselves from food poisoning by cooking and storing food correctly. All meat should be cooked right through and, if re-heated, cooked at a high temperature until piping hot. Cooked food should never be stored in the refrigerator next to uncooked meat. The fridge should always be kept at 2-5°C and the freezer at −18°C. Babies, young children, pregnant women, ill people and the elderly are particularly vulnerable and should not eat eggs when the yolk is still runny.

If symptoms continue for more than a week a doctor may prescribe a course of antibiotics to kill the bacteria that have caused the infection. But it's best to try and avoid antibiotics as they will also kill harmless bacteria that live in our intestines all the time. In fact, these bacteria actually protect us from certain infections.

29

AN IRRITABLE BOWEL

As many as one in three people suffer from an irritable bowel. They have frequent bouts of unexplained diarrhoea or constipation, stomach pain and wind. The problem may go away for a while. But as soon as they are in a stressful situation it comes back.

As well as the basic symptoms, sufferers may also feel tired and anxious and they may experience all sorts of gurglings and rumblings in their stomachs, along with very painful heartburn and frequent indigestion.

For many years doctors thought that people with irritable bowel syndrome were just neurotic, or perhaps hypochondriacs. But they take the problem more seriously now. It seems that some people, especially women, just have more trouble with their digestive systems than others.

There is no miracle cure for irritable bowel syndrome. But eating plenty of fresh fruit and vegetables will give you the fibre you need to help your intestine to digest your food and get rid of waste material as effectively as possible. People with irritable bowel syndrome should avoid fatty and highly processed foods, and drinks such as coffee and tea, that contain caffeine and other stimulants. They should also avoid stressful situations. Or, if that isn't possible, learn how to deal with them. Relaxation classes may be helpful.

People who suffer from irritable bowel syndrome are advised not to drink coffee, tea or cola. The caffeine in these drinks will stimulate their intestines and cause discomfort. Caffeine free coffee, cola and herbal tea are a much better bet.

ABOVE AND BELOW
It's worth doing the revision if you want to feel confident instead of sick when you go into the exam room!

CURING EXAM NERVES

Everyone gets exam nerves at some time during their school days. 'Butterflies in your tummy' is a good description. Your tummy turns over, your heart beats fast, you sweat and you may feel rather sick. As you stand outside the exam room, take some deep breaths and try to think about something you enjoy doing. Shrug your shoulders up and down several times and relax them. Shake your arms about and relax them too. Close your eyes and take some more deep breaths.

This sort of relaxation technique should calm you down and get rid of some of the butterflies. But it's no substitute for revising properly!

THE GREAT BEEF SCARE

Thousands of cows have died from BSE. Nowadays, cattle feed no longer contains bone meal because of the risk of infection.

In 1992, Moscow health officials rejected a large batch of free beef from Britain despite the shortage of meat in the city. They were afraid that it might carry 'mad cow disease' (bovine-spongiform-encephalopathy, BSE).

British cows aren't the only ones to get BSE, but they have had most of the bad publicity. BSE first appeared in 1985 and within two years it was claiming thousands of animals a year. Infected cows become edgy and irritable, unco-ordinated and aggressive and they have to be put down. Their carcasses have to be destroyed and they may not be used as food for humans or animals.

No one knows exactly what causes BSE. Sheep suffer from a related disease, called scrapie. But the only human disease that may have a similar cause is a rare, fatal form of dementia called Creutzfeld Jacob disease.

Scientists may not know what causes BSE, but they believe they know how it got into cattle. Cattle feeds used to contain bone meal produced from other animals and it looks as though these were infected with the organism that causes BSE.

LEFT When you buy a chicken at the supermarket, you don't usually think about what it ate when it was alive. However, animal feed routinely contains chemicals to keep the animals healthy. No one knows what the long-term hazards of eating such meat might be.

BELOW This allergic rash was caused by the drug, penicillin, which the patient took to treat an infection.

What worried the Russians in 1992 – and still worries many people – is that, just as the BSE bug was passed to cattle in their food, so it might pass to humans who eat infected beef.

There is no evidence that BSE has passed to humans, but any farmer who sold an infected animal for human consumption would be heavily fined. However, farmers only know their animal is infected when it displays symptoms, and animals may be sold for meat before anyone realizes they are infected. This may sound worrying. However, scrapie has been a problem in sheep for many years and there is no evidence that we have been infected by eating lamb or mutton.

A lot of research is going into BSE. Hopefully, scientists will soon have some of the answers.

THE THINGS ANIMALS EAT

In 1992 a British woman who had been on holiday in Tenerife went to her doctor with all the symptoms of an allergy to the antibiotic, penicillin. She had a bad rash and her face, throat and limbs were swollen. Yet, she hadn't been taking antibiotics.

Doctors discovered that, while on holiday, she had eaten chicken that was likely to have been fed penicillin and that's why she had come out in a rash.

Some people believe that agricultural animals are fed far too many things that could be dangerous to us if passed on through meat, milk or eggs. Feeds routinely contain drugs, such as antibiotics, to keep animals healthy. Some include chemicals to make them grow better too. No one knows what the long-term hazards might be.

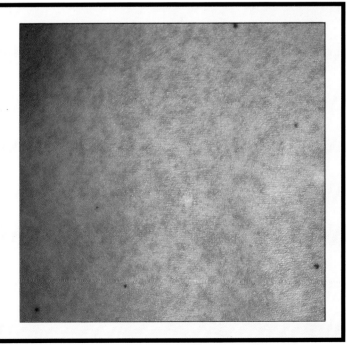

CANCER

Fibre helps to speed up digestion which may reduce the risk of cancer.

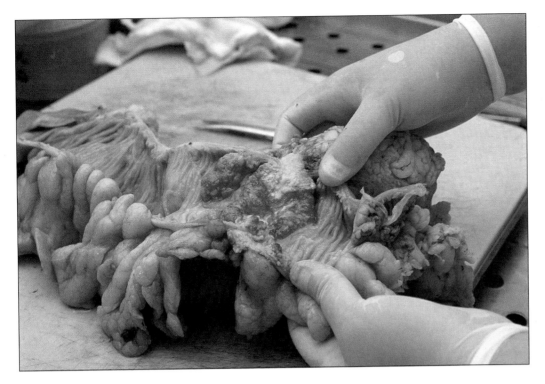

A pathologist examines a human colon containing a cancerous tumour. The person this colon came from is still alive as the cancer was removed before it spread around his body.

Cancer can attack any part of the digestive system. But it is most likely to affect the colon or rectum. If a tumour is discovered when it is very small, there is an excellent chance of curing it. But, unfortunately, it is very difficult to find a tumour in the colon or rectum before it has started to spread to other parts of the body, such as the liver.

People with cancer rarely die from the first tumour that grows in their bodies. They usually die from the secondary tumours, called metastases, that spread to vital organs.

This is why it is so important to recognize the earliest symptoms. Someone with a tumour in their colon or rectum may notice blood in their faeces when they go to the toilet. They may feel very tired and a blood test will show they are anaemic. They may not want to eat and so they lose a lot of weight.

Of course, there are lots of reasons why people feel tired or lose weight. And they may see some blood when they go to the toilet because they have been straining too hard. But if the symptoms continue, it's best to see a doctor.

No one knows exactly what causes cancer of the colon or rectum. You cannot 'catch' cancer; it is not infectious. But this kind of cancer is less common in countries where people eat a lot of vegetables and fruit. It is more common where people eat a lot of rich, processed foods.

Experts think that it is the fibre in vegetables and fruit that may protect people. Fibre helps to speed up digestion. So there is less time for any harmful chemicals in our food to come into contact with the delicate lining of the bowel.

Stomach cancer has also been linked to what we eat. It's the only common cancer that is actually becoming more unusual. That may be because food is stored more carefully than it used to be and we eat less salted and smoked foods.

CANCER FAMILIES

Tumours of the colon and rectum are more common in some families than others. Some people seem to inherit a tendency to get the disease from their parents. This is why special cancer family clinics are being set up at big hospitals to keep a check on families that are at risk. They can have regular tests for tumours so they can be discovered earlier and treated more effectively.

THE DANGERS OF ALCOHOL

You can't tell how much alcohol a drink contains just by looking at it! Although these three drinks are all different sizes, they all contain one unit of alcohol.

One in four men and one in twelve women drink too much alcohol. Heavy drinking over a long period of time is bad for the liver, the mouth, the oesophagus, the larynx, the heart and the brain. It can also cause personal problems and make men unable to have children.

Doctors recommend that men drink a maximum of twenty-one units of alcohol a week and women drink a maximum of fourteen units. One unit of alcohol is equal to one glass of wine or 0.5 litres of ordinary strength beer or lager, a measure of spirits, such as gin or vodka, or a small sherry.

A woman should drink less alcohol because her body contains more fat and less water than a man's. So alcohol is more concentrated in a woman's body and therefore does more damage.

A woman's stomach also breaks down alcohol less efficiently than a man's. So when it reaches the liver it is more likely to kill cells.

The liver's job is to break down unwanted substances in the body, ready to be excreted. That includes alcohol. If you drink too much alcohol, the chemicals in it will destroy the cells of the liver so it can no longer process the body's waste. This means that toxic chemicals will build up in the blood and this can be fatal.

Alcohol is not fussy about which cells it kills. The brain is probably the most vulnerable part of the body. If you kill brain cells with alcohol you won't grow new ones.

Doctors now agree that the odd glass of wine or lager won't do most adults any harm. But people who drink much more than this can do themselves a lot of harm.

SOME FRIGHTENING STATISTICS

Alcohol can be fatal in other ways too. About a quarter of thirteen to seventeen year olds get into arguments or fights after drinking. One in five people who drown have been drinking. Over a third of drivers aged twenty to twenty-four who are killed in road accidents have more than the legal limit of alcohol in their blood. Heavy drinking is a factor in 20-25 per cent of cases of child abuse.

POISONING

A poison is any substance that can make you ill. Not just cyanide and arsenic, but 'hard drugs' such as heroin and cocaine. Even tobacco, alcohol, household chemicals, and certain drugs that you can get from the doctor can be classed as poisons.

Once inside your body it is very hard to stop a poison from harming you. This is because it will go through your digestive system and find its way into your bloodstream, just like anything else you eat or drink.

Each year, thousands of children and adults are taken to hospital because they have swallowed something poisonous. Some do it accidentally, others on purpose. Some die.

Several poisons stop the lungs from working so you can't breathe. Others make the heart stop beating or the blood pressure go dangerously low. Many poisons will also damage the liver and kidneys since these are the organs that deal with them before they are excreted.

Some poisons work very quickly, others gradually build up in the body and take weeks or months to kill. What is safe for one person may be very dangerous for another. This is very important for medicines. A dose of a drug that can save one person's life may be toxic to someone else. For example, children and elderly people need much lower doses of most medicines than healthy adults.

Children under twelve should not take the painkiller aspirin, or any medicines containing aspirin. In this age group, it can cause severe brain and liver damage which may even be fatal.

STAYING SAFE

Never eat berries or plants you see when you are walking in the country unless you are certain you know what they are. Wild strawberries and blackberries are delicious. But be sure you don't mistake other berries for them. Also be very careful about eating wild mushrooms. There are lots of different types and only some are safe to eat.

At home, never leave household chemicals, such as bleach, or medicines, where a young child can reach them. You may know they are dangerous but your little brother or sister may not.

LEFT one of these glasses contains lemon squash, the other contains bleach. A child could drink the bleach instead of the squash and be badly injured. Always keep household chemicals out of reach of small children.

RIGHT This toadstool may be beautiful, but in fact, it is extremely poisonous.

SWEET OR SAVOURY?

Do you have a 'sweet tooth'? Do you always choose sweet, sugary foods like cakes rather than savoury ones like crisps?

Your teeth have nothing to do with the food you choose. They would probably prefer that you kept away from sweet things! It's your tongue that helps decide what sort of food you like. Your tongue contains taste buds. There are about 9,000 of them on an average adult tongue. Look in a mirror and stick out your tongue and you'll see them! Taste buds are goblet-shaped structures that contain cells which can detect different tastes.

There are four taste zones. On the tip of your tongue the taste buds detect sweetness. Just behind them on the sides of your tongue, the taste buds pick up salty flavours. Further back, but still on the sides, are the sour taste buds. And, at the very back of your tongue, are the taste buds that register bitter tastes.

When you put something into your mouth it will pass over the taste buds on your tongue. If it is a sweet food it will stimulate the sweet taste buds and they will send messages down nerves to the brain. The messages will be decoded and, within a second or two of starting to eat, you will know that the food is sweet. A similar system works for the other types of taste buds.

Smell also plays an important part in telling you what sort of food you are eating. If, for example, you pinched your nose and shut your eyes you probably wouldn't be able to tell the difference between potatoes, apples and onions.

People vary in the way they recognize taste too. Some may need a lot more of one type of food to trigger their taste buds than others. And certain tastes are inherited. For example, 60 per cent of the population can taste the bitter chemical, phenylthiocarbamide, and 40 per cent cannot.

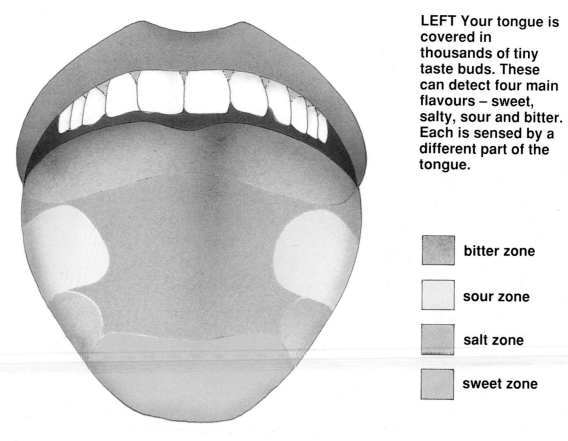

LEFT Your tongue is covered in thousands of tiny taste buds. These can detect four main flavours – sweet, salty, sour and bitter. Each is sensed by a different part of the tongue.

▪ bitter zone

▪ sour zone

▪ salt zone

▪ sweet zone

See if you can tell the difference between onion, cheese and apple when you pinch your nose and shut your eyes. It's not easy!

Most children prefer sweet foods. It's not surprising really as foods hit the 'sweet' taste buds on the tip of the tongue first. Even children who are not given sweets and biscuits usually develop a 'sweet tooth' when they do taste them.

In general, people acquire more of a taste for savoury foods as they get older. An adult's idea of a 'treat' tends to be going out for a meal rather than eating sweets. But, once tasted, few ever lose their love of chocolate!

ADDING UP THE CALORIES

Most sweet foods contain more calories than savoury ones. Compare the 578 calories in 100 g of delicious milk chocolate with the 8 calories in the same amount of lettuce! But some foods can catch you out. For example, 100 g of roasted peanuts – which most of us would find quite salty – contain 586 kilocalories, and 100 g of almonds a few calories less.

Luckily, scientists have discovered several ways of helping us out. For example, artificial sweeteners, such as saccharin or aspartame, can be used to sweeten fizzy drinks or tea and coffee (right).

TOOTH DECAY

How many fillings do you have? Children today have a fraction of the fillings that their parents had. You can thank the fluoride in your toothpaste for saving you from the dentist's drill. Fluoride makes teeth more able to resist decay.

Dental diseases fall into two groups – those that affect the teeth and those that attack the gums. Indirectly, gum disease also affects your teeth because if the gums are diseased your teeth may fall out.

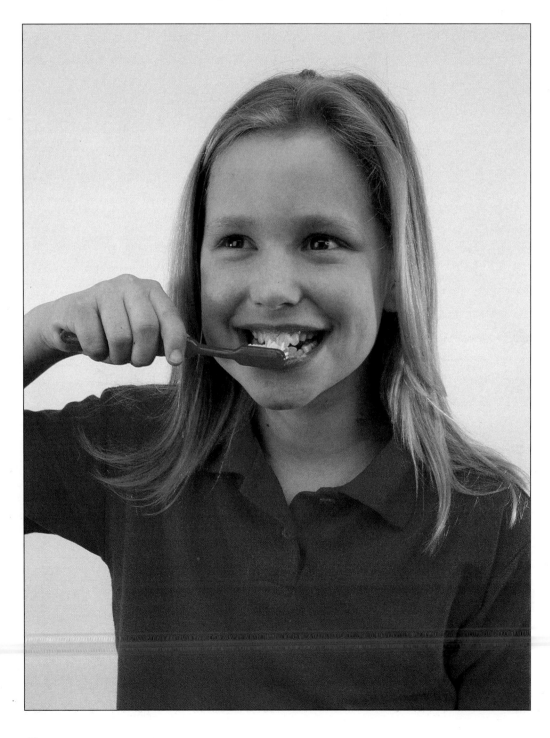

Brushing teeth regularly with a fluoride toothpaste helps keep teeth free of decay. The fluoride is absorbed into parts of the teeth which are attacked by acid, and makes them strong.

Tooth decay occurs when the bacteria on your teeth convert sugar from your food into acid. This attacks the protective layer of enamel which covers each tooth. When the enamel is gone, the acid can go deeper and deeper into the tooth until you are left with blackened pegs instead of pearly-white, healthy teeth.

Bacteria can only set off this horrible process if you let them. They live in the creamy, smelly plaque that builds up on your teeth if you don't brush them. Plaque is a mixture of saliva, the remains of food and bacteria.

Chemicals in saliva can harden the plaque on your teeth into deposits, called calculus. You cannot brush these off; only the dentist or hygienist can get them off. Calculus on your teeth tells your dentist more about your brushing habits than you would ever admit. It's a sure sign that you are heading for tooth decay.

Bacteria in plaque can also cause gum disease. The gums become inflamed and bleed. The gums later recede and the inflammation spreads to the roots of the teeth. The fibres that anchor them in place come apart and eventually the teeth loosen and fall out.

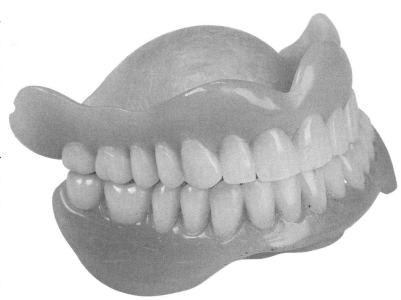

ABOVE Not very pretty are they? This is what you will need if you don't look after your teeth.

CHEWING GUM

Chewing gum after meals can help protect your teeth. But you must choose a gum that is low in sugar. Best of all, choose a gum that contains artificial sweetener, not sugar. The chewing action produces saliva which helps to neutralize the acid in your mouth that causes tooth decay. Tooth decay is one of the major causes of bad breath. So chewing gum, may also make your breath smell better.

FUTURE FOOD

How would you feel about swallowing a handful of tablets three times a day instead of eating breakfast, lunch and tea? Not much fun? It's hard to imagine that we'll ever find ourselves taking this type of space-age nutrient.

But our dietary habits are changing. We are eating less meat, especially beef and lamb, and more chicken. In fact, more and more people are becoming vegetarian.

We expect a much wider range of foods in the shops, all year round. Only twenty years ago, people ate vegetables such as cabbage, carrots and old potatoes in winter, and salad and new potatoes in summer. Mangoes, kiwi fruit, even pineapples were exotic foreign fruits which were rarely found outside the countries where they were grown.

Today, you can get all these foods, summer and winter alike. That's because, thanks to more widespread use of greenhouses, they are grown in a lot more countries and can be transported more easily. Also many foods can be kept for longer before they go off. Either they are stored better or, in some cases, they are irradiated or processed to make them keep longer.

Genetic engineering is even being used to make fruit and vegetables stay fresh. Scientists have already changed the genes of tomatoes so that they don't go mushy as quickly. Animals are also being changed genetically to grow more quickly, to produce leaner meat (less fat) and, in some cases, to make human hormones that are needed for some kinds of medical treatment.

What would happen to all the expensive restaurants if we ate meals like this?

Organic vegetables are usually more expensive than other vegetables. But they are grown without pesticides and fertilizers. They may not look as appetizing though.

How far should the scientists go? Governments keep a close check on the experiments being carried out as the long-term effects of putting new genes into plants and animals are not known.

It is possible that serious problems will come to light. But if all goes well, we could soon be buying the perfect chicken, the tastiest tomato, the juiciest orange and the leanest steak – every time.

Non-organic vegetables are cheaper and look better than organically-grown produce. But do we want all our tomatoes, cauliflowers, carrots and potatoes to look the same?

GLOSSARY

allergy unusual reaction to something you breathe, eat or touch. For example, coming out in a rash after eating strawberries.

amino acids molecules that make up proteins.

anorexia eating problem; people starve themselves because they think they are too fat.

antibiotic type of drug that can kill bacteria and is used to treat infections.

antibody protein which is produced in the blood to fight harmful bacteria.

anus opening at the end of the intestines through which waste products (faeces) are excreted.

bovine spongiform encephalopathy (BSE) also called 'mad cow disease'. A disease which makes cattle unsteady and aggressive. No one knows what causes it.

bulimia eating problem; people eat huge amounts of food and then make themselves sick so as not to put on weight.

caecum part of digestive system between the small and large intestine.

calculus hardened substance which forms on teeth if they aren't cleaned properly. It can cause tooth decay.

calorie measure of energy in food; strictly speaking 'kilocalorie'.

cell very small unit of animal organs and tissues. Place in the body where food is finally broken down to give energy.

colon longest part of the large intestine.

diabetes disease in which people cannot control the amount of sugar in their blood because they have too little of a hormone called insulin.

duodenum first and smaller part of the small intestine where food goes after it has gone through the stomach.

enzyme substance which is needed for a chemical reaction to take place in the cell.

gall bladder small 'bag' of tissue used to store bile before it goes into the intestine.

gene piece of DNA that is responsible for things like the colour of our eyes, the colour of our hair and whether we are tall or small. These are called physical characteristics.

genetic engineering changing animal or human genes in the laboratory.

hypochondriac someone who thinks they are ill when they are not.

ileum main part of the small intestine where digestion takes place.

irradiate to treat with light.

oesophagus tube that carries food from the mouth to the stomach.

metastasis spread of cancer tumours from one part of the body to others.

plaque creamy material which collects on the teeth, made up of bacteria, food particles and saliva. Causes tooth decay.

rectum lower end of the large intestine where faeces are stored before being excreted.

BOOKS TO READ

The Amazing Voyage of the Cucumber Sandwich by Dr Peter Rowan (Jonathan Cape Ltd, 1991)

The Human Machine by Brenda Walpole (Wayland, 1990)

Diet and Nutrition by Brian R. Ward (Franklin Watts, 1987)

Let's Discuss Health and Fitness by Tony Wheatley (Wayland, 1988)

Your Body Fuel by Dorothy Baldwin & Claire Lister (Wayland, 1983)

The Body and How it Works by Steve Parker (Dorling Kindersley, 1987)

Twentieth Century Medicine by Jenny Bryan (Wayland, 1988)

Pocket Book of the Human Body by Brenda Walpole (Kingfisher Books, 1987)

ACKNOWLEDGEMENTS

Action Plus 13 (Peter Tarry); Aquarius 16: Chapel Studios cover and title page, 6, 15 (top), 21 (bottom), 29 (bottom), 30, 35 (top), 37, 38, 41 (top and bottom), 42, 43 (bottom), 44, 45 (top and bottom); Eye Ubiquitous 11, 15 (bottom), 27 (Ken Oldroyd) 31 (top and bottom); Life Science Images cover background, 25 (top), 34; Peter Sanders 17; Sally and Richard Greenhill 20; Science Photo Library 4 (Marcelo Brodsky), 5 (Tony Craddock), 9, 11 (Tony Craddock), 18 (Edward Lettaw), 23 (top CNRI/bottom A.B. Dowsett), 26 (Chris Priest and Mark Clarke), 28 (A.B. Dowsett), 33 (bottom, Dr J. Burgess); Sparks Photographics 39; St Bartholomew's Hospital 8, 19; Topham 33 (top); W.P.L. 10, 21 (top), 29 (top), 36; Zefa 7, 12, 22, 24, 32, 43 (top). Artwork by Malcolm Walker.

INDEX

alcohol 36-7
allergy 24-5,33
amino acid 7
anaemia 19
anorexia 20-21
antibiotics 23
antibodies 24,33
appetite 15,20
anus 6
arthritis 12
aspirin 38
asthma 25

bile 8
bulimia 20-21

caecum 6
calorie 10
cancer 4,12,35
carbohydrate 6,8,15
cells 6,22-3,27
colon 6,35
Creutzfeld Jacob disease 32

dairy products 7,25
diabetes 9,12,26-7
diaphragm 8
diarrhoea 4,21,28-30
diet 12,14-15,20
    balanced 10
duodenum 6,8,22

energy 7,15,19
enzymes 6,8

excrete 6,8
exercise 14

faeces 6
famine 19
fasting 17
fat 6,8,10,15
fibre 10,30,33

gall bladder 8
genetic engineering 44
glucose 15,26-7
greenhouses 44
growth 7

hay fever 25
heartburn 22,30
hepatitis 28
histamine 23
hormone 8,26
hydrochloric acid 6

ileum 6
indigestion 22,30
intestine 4,6,8
intolerance 24-5
insulin 8,27
irradiated 44
Islets of Langerhans 8-9

kidneys 38

laxative 21
liver 8-9,37-8

mad cow disease (BSE) 32-3
malnutrition 18-19
migraine 25
mineral 7,10

nutrient 4

oesophagus 6

pancreas 8-9,27
poison 38
processed 4
proteins 6,8,10

Ramadan 17
rectum 6,35

salmonella 28
scurvy 19
stomach 6,22-3

taste buds 40
teeth 10,42-3
thin 18,20
tongue 40-41

ulcers 4,22-3

vegetarian 10
vitamin 6,7,10
vomiting 4,28-9

weight 10,12-14,27